T0403307

THE ECLIPSE OF 1919

How Einstein's Theory of General Relativity Changed Our World

THE
ECLIPSE
OF 1919

How Einstein's Theory of General Relativity Changed Our World

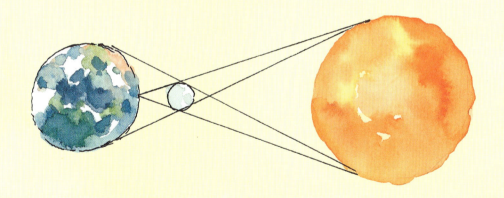

EMILY ARNOLD McCULLY

Christy Ottaviano Books
Little, Brown and Company
New York Boston

PROLOGUE

Why do the planets stay in their orbits around the sun? Why does the moon circle the earth and not fly off into space? What makes an apple—or any other object—fall to the ground?

In 1687, Isaac Newton answered these questions. He said gravity is a force that holds the planets in their orbits, pulls the apple toward Earth, and acts on all objects everywhere in the universe.

Newton's idea stood for two hundred years. But it didn't explain everything. Newton didn't know that light has a speed limit of about 186,282 miles per second and that nothing in the universe can travel faster. If that was true, how could the sun instantly move a distant planet when sunlight took hours to reach it? And why does a heavy object fall to the earth at the same rate as a lighter one?

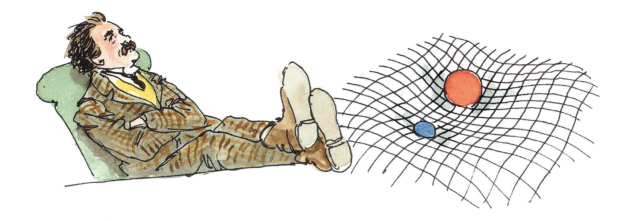

Albert Einstein used thought experiments to answer these questions. He concluded that energy and mass are interchangeable ($E = mc^2$) and imagined that distance (or space) and time are really two different ways to look at the same thing. He called them space-time. Einstein imagined gravity as masses (objects) warping (or bending) the space-time around them. Even energy, such as beams of light, can bend space-time. Speed affects mass, time, and space. He called that relativity.

These were wild ideas. Hardly anyone understood them. Many scientists insisted that Newton's universal law of gravity was correct and Einstein's was wrong. They believed what their eyes could see. An apple seems to be pulled in a straight line to the ground. You can't see an apple's curved space-time.

Like any theory, what Einstein called general relativity had to be tested before it was accepted. Einstein thought he knew a way to test his theory. What would happen in such a test? The world learned the answer during a total eclipse of the sun in 1919.

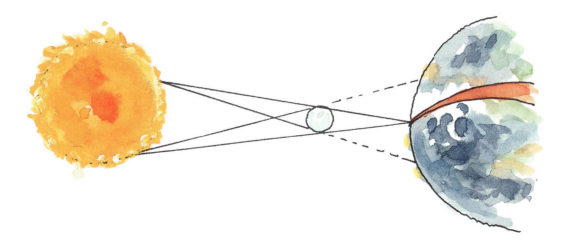

YOUNG ALBERT EINSTEIN

Even as a toddler living with his family in Germany, Albert Einstein seemed lost in his thoughts. The family's maid called him "the dopey one."

Once, when he was five and sick in bed, his father gave him a compass. Albert was fascinated. No matter how he held it, the needle always pointed north. Some invisible force moved it. But what? He got goose bumps trying to understand it. All kinds of things that seemed ordinary to other people were, for Albert, mysteries to be solved.

School did not go well for Albert. He hated being taught only what people already knew. Albert wanted to find new answers to big questions. Practicing his violin helped him to think.

One morning, daydreaming as usual, Albert wondered what would happen if he could run fast enough to overtake a beam of light.

Wouldn't the light seem to stand still?
That was supposed to be impossible. Wrestling
with this problem made his palms sweat.

From Earth, it would look as if
he could catch the light beam.

After an unspectacular academic career in high school, Albert entered a technical college in Switzerland to study physics. His professors didn't think much of him. He cut classes and did his serious thinking on his own.

Albert graduated fourth in his class of five and went looking for a job. Everyone turned him down. A friend recommended him to the Swiss Patent Office in Bern. The pay was good, and reviewing patent applications would give him time for his own thoughts. He was hired.

Whenever he had a moment, Albert wrestled with ideas like racing the beam of light. He and his friends argued over them while hiking in the mountains. He liked to conduct thought experiments in a sailboat on the lake.

Gradually, Albert figured out a radical new way to describe the universe, beginning with how motion and time, energy and mass, all work together. He believed they behaved in relation to one another. The key to the universe was relativity.

EARTH

EINSTEIN AT WORK

Albert Einstein knew relativity was beautiful. It had elegance and simplicity. But was it correct? The effects he described can't be observed in the everyday world because everything is closer together than in space.

STAR IS HERE.

STAR APPEARS
TO BE HERE.

Newton had predicted that the light from a star would be bent by the force of the sun's gravity. If Einstein was right and gravity was the effect of the object's curved space-time, the star's light would bend twice as much, making it look as if the star itself had moved.

Was Einstein right? Because you can't look at stars that seem to be near the sun in daytime, his theory could be tested only during a total eclipse.

The next total solar eclipse was expected on August 21, 1914. Einstein asked an astronomer friend, Erwin Finlay-Freundlich, to photograph it from a location in Russia. But then World War I began.

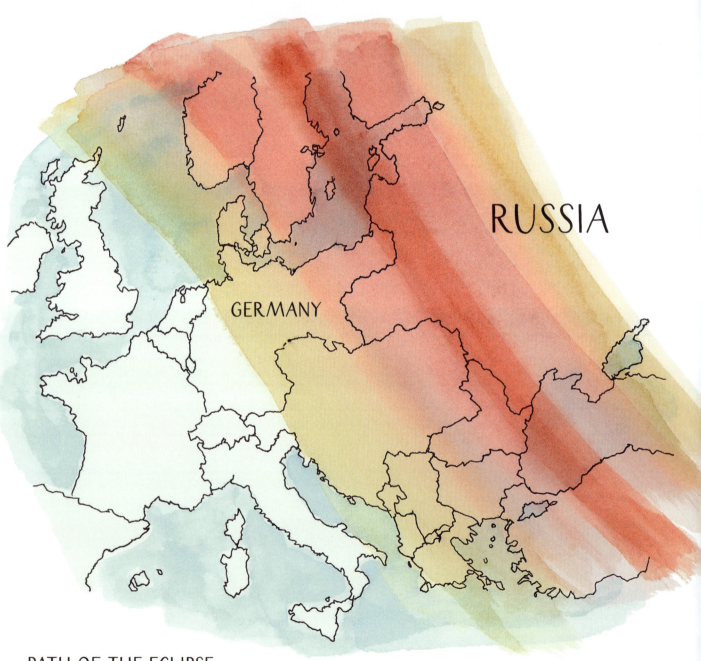

RUSSIA

GERMANY

PATH OF THE ECLIPSE
AUGUST 21, 1914

When Freundlich tried to enter Russia, which was at war with Germany, he was arrested and his equipment was seized. That ended the experiment to test relativity. As it turned out, it was a good thing: Einstein reviewed his calculations and discovered that he'd made a mistake. The experiment would have disproved his theory.

That was also the year Einstein was finally offered a job worthy of his brilliance, at the University of Berlin. His reputation was slowly spreading. Then, in 1915, he published his theory of general relativity, with its revolutionary description of the relativity of time and motion and the warping of space-time by gravity.

World War I was raging in Europe, and Great Britain and Germany were on opposite sides. German scientists were developing horrific weapons like poison gas. Einstein's colleagues in Berlin shunned him because he opposed all war. In Britain, scientists wanted nothing to do with ideas from Germany. Almost nobody paid attention to Einstein's new theories.

But one Englishman, Arthur Eddington, director of the Cambridge Observatory, received a copy of Einstein's paper from a Dutch physicist. Eddington understood it. Like Einstein, he hated war and refused to join the army. A Quaker pacifist, he faced imprisonment.

Sir Frank Dyson, England's Astronomer Royal, devised a plan to keep Eddington out of jail while advancing science. Eddington would test the theory of relativity by photographing an eclipse to see if the sun's gravity bent light from a nearby star. The next total eclipse would take place in May 1919.

TESTING THE THEORY OF RELATIVITY

World War I had ended just a few months earlier when Eddington and his assistant, Edwin Cottingham, an expert clockmaker in charge of the measuring instruments, sailed to the island of Príncipe, off the west coast of Africa.

They made camp at a plantation five hundred feet above the sea, set up the telescope and camera, and waited.

On the morning of May 29, 1919, a heavy rain fell. Eddington kept his camera ready, but as the hour approached, the eclipse was covered by thick clouds. The men were heartsick.

Then, at two p.m., just as the moon started to pass between
Earth and the sun, the rain stopped and the clouds parted.

As soon as the sun was covered by the moon's shadow, Eddington began taking photographs, changing the heavy glass plates as fast as he could. It was a long eclipse.

Six minutes later, the clouds returned.

Eddington and Cottingham developed twelve of the sixteen photographic plates on Príncipe. Most were too blurry. But one was not: It clearly showed—compared to a photo taken before the eclipse—that a star seemed to have moved.

Eddington called it the greatest moment of his life. He telegraphed the results to Dyson in code: "Through cloud. Hopeful."

In September Einstein learned that the experiment had succeeded. He sent his mother an understated postcard: "Dear Mother, Today some happy news . . ."

Einstein had never doubted the theory. He celebrated by buying himself a new violin.

The Royal Society of London called a special joint meeting on November 6, 1919. Reporters were invited, but the press didn't expect much.

Some scientists had no idea what was to be announced that day. Many of those who did firmly believed in Newton's idea of gravity and were sure that Einstein was wrong about relativity.

Excitement built as the crowd waited for the Astronomer Royal to speak. When he told the crowd that Eddington had proved Einstein was right, angry Newtonians marched out of the hall in protest.

But the universe *really had* changed. Newspapers around the world ran bold headlines.

LIGHTS ALL ASKEW IN THE HEAVENS

DON'T WORRY OVER NEW LIGHT THEORY

Men of Science More or Less Agog Over Results of Eclipse Observations.

EINSTEIN THEORY TRIUMPHS

Stars Not Where They Seemed or Were Calculated to be, but Nobody Need Worry.

SUN'S GRAVITY BENDS STARLIGHT
Einstein's Theory Triumphs

ECLIPSE SHOWED GRAVITY VARIATION

Diversion of Light Rays Accepted as Affecting Newton's Principles.

HAILED AS EPOCHMAKING

British Scientist Calls the Discovery One of the Greatest of Human Achievements.

REVOLUTION IN SCIENCE

NEW THEORY OF THE UNIVERSE.

NEWTONIAN IDEAS OVERTHROWN.

It Discards Absolute Time and Space, Recognizing Them Only as Related to Moving Systems.

IMPROVES ON NEWTON

A BOOK FOR 12 WISE MEN

No More in All the World Could Comprehend It, Said Einstein When his Daring Publishers Accepted It

WORLD FAMOUS

Overnight, Albert Einstein's name became a permanent synonym for *genius*. The war-weary world was brought together in its astonishment. Almost nobody understood it yet, but relativity would eventually help explain how the universe began (the big bang theory) and where it was going (expanding). Black holes would be imagined and eventually confirmed. Dark matter and dark energy would be discovered and found to make up most of the universe.

Einstein had realized a new way to describe how an uncertain and unpredictable universe works.

And more than a century later there is still plenty to discover!

AUTHOR'S NOTE

Albert Einstein (1879–1955) lived and worked for nearly twenty years in Germany, winning the Nobel Prize in Physics in 1921 (not for the theory of general relativity but for his paper on the photoelectric effect). The rise of Hitler and the Nazi Party finally drove him to settle in the United States in 1933. He worked at the Institute for Advanced Study at Princeton University until his death. Americans loved him for his wild hair and impish sense of humor. He was devoted to peace, civil rights, and his violin. Scientists are still finding ways to test relativity, but no one has proved it wrong.

The theory of relativity predicted gravitational waves and black holes. Einstein's ideas led to such life-changing fixtures as television and GPS. They also contributed, to his great regret, to the development of the atomic bomb.

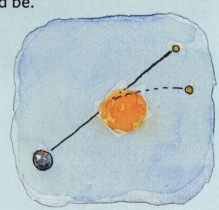

Sir Frank Dyson, the Astronomer Royal, sent a second expedition to Sobral, Brazil, to photograph the 1919 eclipse, and of course it confirmed Eddington's results.

The eclipse made Arthur Eddington (1882–1944) famous too. He is generally known as a founder of astrophysics. At the age of four, he tried to count the visible stars. He was a superb student and, like Einstein, an avid cyclist. For most of his career, he taught at Cambridge University in England and directed the observatory. Someone once said to him that he must be one of only three people in the world who understood relativity. Eddington was silent for a minute. "You don't agree?" the man asked. "Oh," said Eddington, "I was just wondering who the third person could be."

Eddington composed a bit of light verse about the eclipse:

"Oh leave the Wise our measures to collate
One thing is at least certain, LIGHT has WEIGHT,
One thing is certain, and the rest debate—
Light-rays, when near the Sun, DO NOT GO STRAIGHT."

To photograph the eclipse, Eddington had to slide large, fragile glass plates into the telescope-equipped camera, then quickly remove and replace them. Reverse images, or negatives, appeared on the coated plates and were developed using cold water. Eddington was able to develop them on the tropical island of Príncipe, where the available water for the developing fluid was warm, because he had access to ice cubes!

ECLIPSE OF THE SUN

Total eclipses of the sun are spectacular events that used to incite alarm before people understood what was happening. They occur about twice a year somewhere on Earth but are seen over only a small area. The moon's shadow usually passes either above or below the earth, but during an eclipse, the earth, moon, and sun all align and the moon's shadow covers the sun, ending the day for about two minutes before it begins again. The Great American Eclipse on August 21, 2017, was seen from Seattle to Charleston. Americans experienced another on April 8, 2024, when it passed diagonally over the United States from Texas to New England. The next total eclipse over part of the contiguous United States is predicted for August 23, 2044.

SOURCES

Dine, Michael. "How Einstein Arrived at His Theory of General Relativity." Literary Hub, February 10, 2022. https://lithub.com/how-einstein-arrived-at-his-theory-of-general-relativity/.

Eddington, Sir Arthur Stanley. *Space, Time and Gravitation: An Outline of the General Relativity Theory.* Cambridge: Cambridge University Press, 1920; repr., 1999.

"Einstein Thought Experiments." Nova Online, September 8, 1997. Includes animated video with narration. 3 min., 12 sec. https://www.pbs.org/wgbh/nova/video/einstein-thought-experiments/.

"Einstein's Theory of Relativity Explained in One of the Earliest Science Films Ever Made (1923)." Open Culture, May 25, 2018. Includes *The Einstein Theory of Relativity.* Produced by Fleischer Studios (Max and David Fleischer), 1923. 20 min., 11 sec. http://www.openculture.com/2018/05/einsteins-theory-of-relativity-explained-in-one-of-the-earliest-science-films-ever-made-1923.html.

Gates, S. James, Jr., and Cathie Pelletier. *Proving Einstein Right: The Daring Expeditions That Changed How We Look at the Universe. Audiobook.* New York: Hachette Book Group; Ashland, OR: Blackstone Audio, 2019.

Gilpin, Caroline Crosson. "Teaching Activities for: 'How to Watch a Solar Eclipse.'" *New York Times,* August 16, 2017. https://www.nytimes.com/2017/08/16/learning/teaching-activities-for-how-to-watch-a-solar-eclipse.html.

Greene, Brian. "Special Relativity in a Nutshell." Nova Online, September 15, 2011. https://www.pbs.org/wgbh/nova/article/special-relativity-nutshell/.

Isaacson, Walter. *Einstein: His Life and Universe*. New York: Simon & Schuster, 2007.

Kennefick, Daniel. *No Shadow of a Doubt: The 1919 Eclipse That Confirmed Einstein's Theory of Relativity*. Princeton, NJ: Princeton University Press, 2019.

"Lights All Askew in the Heavens." *New York Times*, November 10, 1919. ProQuest Historical Newspapers, *New York Times* (1851–2007). Reproduced with permission of the copyright owner. Further reproduction prohibited without permission.

Overbye, Dennis. "Does the Universe Still Need Einstein?" *New York Times*, November 19, 2018. https://www.nytimes.com/2018/11/19/science/einstein-physics-universe.html.

Overbye, Dennis. "During an Eclipse, Darkness Falls and Wonder Rises." *New York Times*, August 14, 2017. https://www.nytimes.com/2017/08/14/science/watching-eclipse-august-21.html.

"Physics of Time." Exactly What Is Time? http://www.exactlywhatistime.com/physics-of-time/relativistic-time/.

"Special Relativity / Elementary Tour part 3: The relativity of space and time." Einstein Online. https://www.einstein-online.info/en/relativity_space_time/.

Stanley, Matthew. *Einstein's War: How Relativity Triumphed amid the Vicious Nationalism of World War I*. New York. E. P. Dutton, 2019.

For Ethan McCully

The author would like to thank
Richard Slovak for his factual review of this book.

About This Book
The illustrations for this book were done in watercolor and pen
and ink on Arches paper. This book was edited by Christy Ottaviano and
designed by Angelie Yap. The production was supervised by Kimberly Stella, and
the production editor was Annie McDonnell. The text was set in Ideal Sans and Salt
and Spices Pro, and the display type is Mostra Nuova.

Copyright © 2025 by Emily Arnold McCully • Cover illustration copyright © 2025 by Emily Arnold McCully • Cover design by Angelie Yap • Cover copyright © 2025 by Hachette Book Group, Inc. • Hachette Book Group supports the right to free expression and the value of copyright. The purpose of copyright is to encourage writers and artists to produce the creative works that enrich our culture. • The scanning, uploading, and distribution of this book without permission is a theft of the author's intellectual property. If you would like permission to use material from the book (other than for review purposes), please contact permissions@hbgusa .com. Thank you for your support of the author's rights. • Christy Ottaviano Books • Hachette Book Group • 1290 Avenue of the Americas, New York, NY 10104 • Visit us at LBYR.com • First Edition: July 2025 • Christy Ottaviano Books is an imprint of Little, Brown and Company. • The Christy Ottaviano Books name and logo are registered trademarks of Hachette Book Group, Inc. • The publisher is not responsible for websites (or their content) that are not owned by the publisher. • Little, Brown and Company books may be purchased in bulk for business, educational, or promotional use. For information, please contact your local bookseller or the Hachette Book Group Special Markets Department at special.markets@hbgusa.com. • Library of Congress Cataloging-in-Publication Data • Names: McCully, Emily Arnold, author, illustrator. • Title: The Eclipse of 1919: How Einstein's Theory of General Relativity Changed Our World / by Emily Arnold McCully. • Description: First edition. | New York : Christy Ottaviano Books, Little, Brown Books for Young Readers, 2025. | Includes bibliographical references. | Audience: Ages 5–9 | Summary: "An awe-inspiring picture book biography about Albert Einstein and the story of how he proved his Theory of Relativity during the 1919 solar eclipse." — Provided by publisher. • Identifiers: LCCN 2024004515 | ISBN 9780316475525 (hardcover) • Subjects: LCSH: Einstein, Albert, 1879–1955—Juvenile literature. | Solar eclipses—1919—Juvenile literature. | General relativity (Physics)—Juvenile literature. | Physicists—Biography—Juvenile literature. • Classification: LCC QB544.19 M33 2025 | DDC 530.11072/3—dc23/eng/20240517 • LC record available at https://lccn.loc.gov/2024004515 • ISBN 978-0-316-47552-5 • PRINTED IN DONGGUAN, CHINA • APS • 10 9 8 7 6 5 4 3 2 1